FORCE

Rebecca Woodbury, Ph.D., M.Ed.

Gravitas Publications Inc.

FORCE!

Illustrations: Janet Moneymaker

Copyright © 2024 by Rebecca Woodbury, Ph.D., M.Ed.

Force
ISBN 978-1-950415-19-9

Published by Gravitas Publications Inc.
www.gravitaspublications.com
www.realscience4kids.com

RS4K

Photo Credits: Cover & Title Page: By sorayut, AdobeStock
Page 3: By Photo Tuller, AdobeStock; Page 5: By Katie, AdobeStock;
Page 7: By Louis-Photo, AdobeStock; Page 9: By JenkoAtaman, AdobeStock

How do you pull a

wagon up a hill?

Wagon rides
are fun!

How do you throw a

ball through the air?

Eat lots of cheese first?

How do you squeeze

a marshmallow?

SQUISH!

How do you ride

a bike?

I don't!

To do all these things,

you use **force!**

Force is
important.

Yes!

FORCE

In physics...

Force is any action that changes...

...the **location** of an object,

...the **shape** of an object,

...how **fast or how slowly** an

object is moving. (This is called

the **speed** of an object.)

When you pull a wagon up a hill, you use **force** to change the **location**.

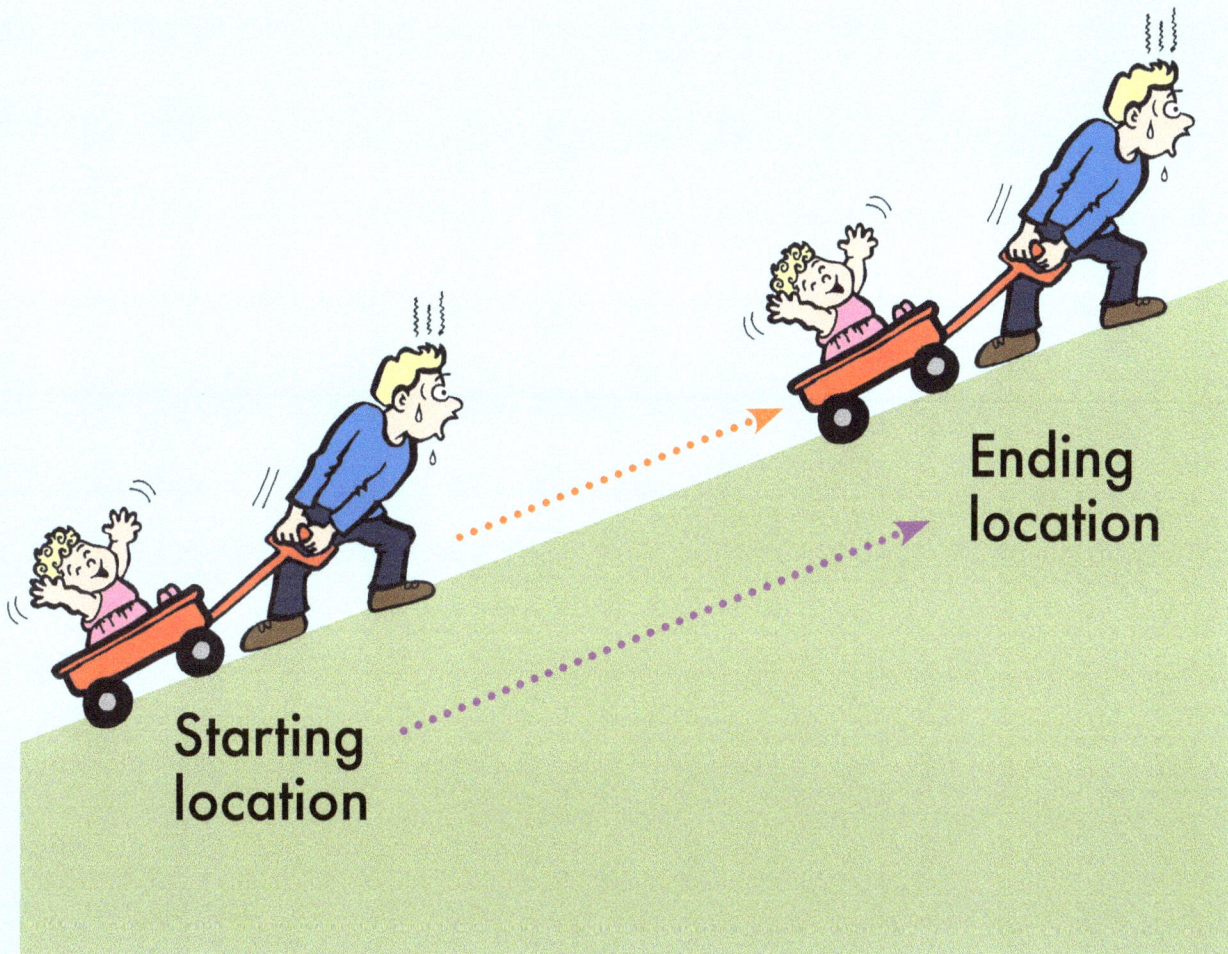

Starting location

Ending location

When you throw a ball through the air, you use **force** to change the **location and the speed**.

Can you throw a ball?

Maybe.

Start ·····················▶ location changes

Start ⟩⟩⟩⟩⟩⟩⟩⟩⟩⟩⟩⟩⟩⟩⟩▶ speed changes

When you squeeze a

marshmallow, you use

force to change the **shape**.

I think he squashed it.

When you ride a bike, you use **force** to change the **speed and location** of the bike and yourself.

Start ·········· → Location changes

Start ///////// → speed changes

Can you tell how **force** is being used?

Summary

In physics...

Force is any action that changes...

...the **location** of an object,

...the **shape** of an object,

...how **fast or how slowly** an object is moving. (This is called the **speed** of an object.)

How to say science words

force   (FAWRSS)

location   (loh-KAY-shun)

object   (AHB-ject)

physics   (FI-ziks)

shape   (SHAYP)

squeeze   (skweez)

www.ingramcontent.com/pod-product-compliance
Lightning Source LLC
Chambersburg PA
CBHW040149200326
41520CB00028B/7549